Applicability of Web 2.0 tools for customer retention

BACHELOR'S PAPER: WEB 2.0 TOOLS THAT CAN CREATE
VALUE FOR COMPANIES

VON TOMÁS PAULITSCH

STUDIENARBEIT

FACHHOCHSCHULE KREMS

Inhalt

4

Kurzfassung

Das Ergebnis dieser Arbeit zeigt, dass Web 2.0 Tools viele Möglichkeiten zur Kundenbindung bieten und besonders die zwei Anwendungen Facebook und Twitter können dafür eingesetzt werden. Dennoch birgt der Einsatz von Web 2.0 Tools in der Marketing Strategie auch immer Gefahren, besonders wenn diese nicht richtig verwendet werden. Am Beispiel von Firmenseiten auf der Social Network Plattform Facebook, sieht man, dass bei falscher Nutzung keine Kohärenz des dort veröffentlichten Inhalts mit dem allgemeinen Firmenimage besteht. Des Weiteren können User sich als Opfer von unerwünschter Werbung fühlen und negative Gefühle gegenüber einem Unternehmen entwickeln. Nichtsdestotrotz bieten Web 2.0 Tools zukunftsorientierten Unternehmen, die wettbewerbsfähig bleiben wollen, großartige Möglichkeiten, wie zum Beispiel aktive Kommunikation mit Kunden, Segmentierung und Zielgruppendefinition, Vertrauens und Akzeptanzförderung, Aufbau von Begeisterung und Emotionen, sowie Erkennen von Trends und Sammeln von Kundendaten. Eine Wertsteigerung für das Unternehmen im Sinne von gesteigerter Brand Awareness, Kundenzufriedenheit, Kundenloyalität, und daraus resultierend, erhöhte Kundenbindung, kann dadurch entstehen.

In einem ersten Schritt wird der Begriff Web 2.0 definiert und beliebte Web 2.0 Tools werden beschrieben und mit praktischen Beispielen ergänzt um die Verständlichkeit des Themas zu erleichtern. Der zweite Teil dieser Arbeit grenzt den Begriff Kundenbindung ein, listet Möglichkeiten und Gefahren auf und beschreibt des Weiteren das Kundenbindungsmanagement. Als letzter Schritt dieser Arbeit werden die zuvor beschriebenen Web 2.0 Tools auf ihre Einsetzbarkeit im Bereich der Kundenbindung analysiert. Der ausführlichste Teil in diesem Schritt beinhaltet den Vergleich der zwei Web 2.0 Anwendungen Facebook und Twitter hinsichtlich ihrer Eignung im Einsatz von verschiedenen zuvor definierten Prozessen der einzelnen Phasen der Kundenbindung. Diese zwei Anwendungen wurden auf Grund ihrer Aktualität und, zur Zeit der Abfassung dieser Arbeit vorhandenen, Beliebtheit gewählt.

Abstract

The outcome of this paper shows that Web 2.0 tools offer many possibilities in order to retain customers and especially the two applications Facebook and Twitter can be used therefor. However, by implementing Web 2.0 tools into a company's marketing strategy also problems can occur, especially if not used with sufficient knowledge. Looking at the example of the social network platform Facebook, one can see that the wrong utilization of corporate pages can lead to inconsistency of the published information and the general company image. Furthermore users can feel over-spammed and thus, can prompt negative feelings for a brand or a company. Nevertheless future-looking companies, which want to stay competitive and differentiate from their competitors, are offered great chances by Web 2.0 tools, such as active communication with customers, segmentation and targeting, building of trust and acceptance, establishing enthusiasm and emotions or identifying trends and collecting customer data. Thus, the usage of Web 2.0 tools can create value for companies, by means of increased brand awareness, customer satisfaction, customer loyalty, and as a result higher customer retention.

In a first step the term Web 2.0 is discussed and furthermore the respective tools are described and their understanding is supported by popular examples of Web 2.0 applications. As a second part this thesis will treat the topic of customer retention, possible opportunities and threats and how to manage it. As a last step the beforehand defined Web 2.0 tools are analyzed according to their applicability for customer retention. The most elaborate part of this step includes the comparison of the two Web 2.0 applications Facebook and Twitter as for their suitability for various beforehand defined processes of the different phases of customer retention. Those two applications were chosen due to recent events and popularity of both, at the time this thesis were written.

1. Introduction

The impact of Web 2.0 on our daily life as well as on our buyer and consumer behavior is of an ever increasing importance. Thus the adoption of Web 2.0 tools is also of growing importance for businesses and advertising companies in order to gain a competitive advantage. Customers can be reached more efficiently and on a broader level. (Aichbauer, 2010, p. 35; Newman, Thomas, 2009, p. 15)

Web 2.0 tools (like Wikis, Blogs, Communities,...) nowadays are used by many customers but should also be introduced by companies in order to enhance internal as well as external communication.

1.1. Description of the problem

Since customer retention is considerably cheaper than the acquisition of new customers, each company should see it as one of its core business operations (Dittrich, 2002, p.16; Mukerjee, 2007, p. 19). Through the growing impact of Web 2.0 customers concurrently win market power because of growing transparence through using Web 2.0 (Duschinski, 2007, p. 52). In order to be able to stay competitive, companies can apply Web 2.0 tools to fulfill important activities needed in customer retention processes.

1.2. Research Questions

The main objective of this thesis is to answer, which Web 2.0 tools can be used in order to fulfill which important activities of customer retention processes. Since Web 2.0 is an area of continuous change two of the currently most successful applications are analyzed as demonstrative examples according to their applicability for customer retention. Hence the following hypothesis can be defined: Web 2.0 tools are applicable for activities needed in customer retention processes, which I expect to be proven right. Furthermore the following research question arises: Are Web 2.0 tools in theory applicable for activities in customer retention processes? Additionally the subsequent questions are subject of discussion:

- What is Web 2.0?
- Which Web 2.0 tools are known and how are they defined?
- What is customer retention and how to manage it?
- Which opportunities and threats are offered by customer retention?
- How can Web 2.0 tools be used in order to successfully manage customer retention?

1.3. Limitations

Focusing on customer retention and how to manage it, limits this thesis in the area of customer relationship management. Hence this paper only discusses the applicability of Web 2.0 tools for the retention of existing customers and not the acquisition of new customers.

Furthermore, since this thesis are only conducted by researching literature and professional articles from the internet, applicability is limited to theory and no practical examples are used. Additionally there is no specification for the different industries and market sectors in which the companies operate in, but the entire thesis is conducted with regards to general applicability of findings. The practical treatment of this topic was not manageable for this short period of research and was furthermore neglected due to a lack of contacts to experts in the respective field.

1.4. Methods

This thesis was conducted through research in literature in the fields of Web 2.0, customer retention, customer relationship management and marketing including English and German books and documents. The information was mainly searched for online or in the IMC library focusing on the keywords aforesaid. Due to the huge amount of literature especially in the field of customer retention, theories and strategies stated in books are sometimes contradictory. During the research process several books and online documents considered as useful for the conduction of this thesis, were in the end disregarded due to better description available or an inconsistent outcome throughout the writing process. Examples in the sector of Web 2.0 tools were chosen as mentioned in respective literature or according to online articles about recent popularity of applications.

1.5. Structure

The paper is divided into three parts:

The first part describes the term Web 2.0 in detail and gives general definitions and facts about Web 2.0 including a detailed description of Web 2.0 tools and which possibilities each tool offers (communication, creation of emotions,...).

The second part especially concentrates on customer retention and the process of managing customer retention in Customer Relationship Management (CRM). Additionally definitions of terms and importance of customer retention and the management of customer retention are stated.

The third part consequently gives some approaches of how to use Web 2.0 for customer retention and furthermore evaluates the Web 2.0 tools, specified in the first part, according to their usability for activities of customer retention processes, described in the second part. For further comprehension and due to recent events, the two Web 2.0 applications Facebook and Twitter are compared and evaluated upon their adequacy in the different customer retention processes.

2. Web 2.0

Defining the term itself and several Web 2.0 tools as well as giving examples of their usage, should give a basic understanding of the topic. The applicability of the chosen tools for customer retention is thereupon comprehensible and important for the final outcome of this paper.

2.1. Definition and history of Web 2.0

The term Web 2.0 evolved in 2004 at a brainstorming session between O'Reilly and MediaLive International, invented in order to describe and pool new and emerging Web-based tools. (Schrum, & Solomon, 2007, p. 13) After the dot-com bust in 2001 the Web was thought to be a thing of the past, though survivors of the burst and newly emerging companies all exploited the new opportunities of the network as a platform. (Shuen, 2008, p. 1)

Among experts there is still "a huge amount of disagreement about just what Web 2.0 means" (O'Reilly radar, 2005-2009), ranging from terms such as "meaningless marketing buzzword" to "new conventional wisdom". (O'Reilly Media Inc., 2011)

In order to clarify the term O'Reilly opposed Web 1.0 applications to those of Web 2.0, concentrating on the principles of conventional and new Web 2.0 tools. The ongoing list is summarized in table 1, specifying some examples of Web 1.0 applications and their new exemplary opponents for Web 2.0. (O'Reilly Media Inc., 2011)

Table 1: Web 1.0 vs. Web 2.0 Applications

Web 1.0		Web 2.0
Double Click	→	Google AdSense
Ofoto	→	Flickr
Akamai	→	BitTorrent
mp3.com	→	Napster
Britannica Online	→	Wikipedia
Personal websites	→	Blogging
Evite	→	upcoming.org and EVDB
Domain name speculation	→	search engine optimization
page views	→	Cost per click
Screen scraping	→	Web services
Publishing	→	Participation
Content management systems	→	Wikis
Directories (taxonomy)	→	Tagging ("folksonomy")
Stickiness	→	Syndication

Source: adapted from: O'Reilly Media Inc., 2011

To show the complexity of the principles and practices around the gravitational core of the Web as a platform, a "Meme Map" was set up at one of the brainstorming sessions. This map, shown in figure 1, is still seen as a work in progress showing up the main ideas emanating from the Web 2.0 core. (O'Reilly Media Inc., 2011)

Figure 1: Meme Map of Web 2.0

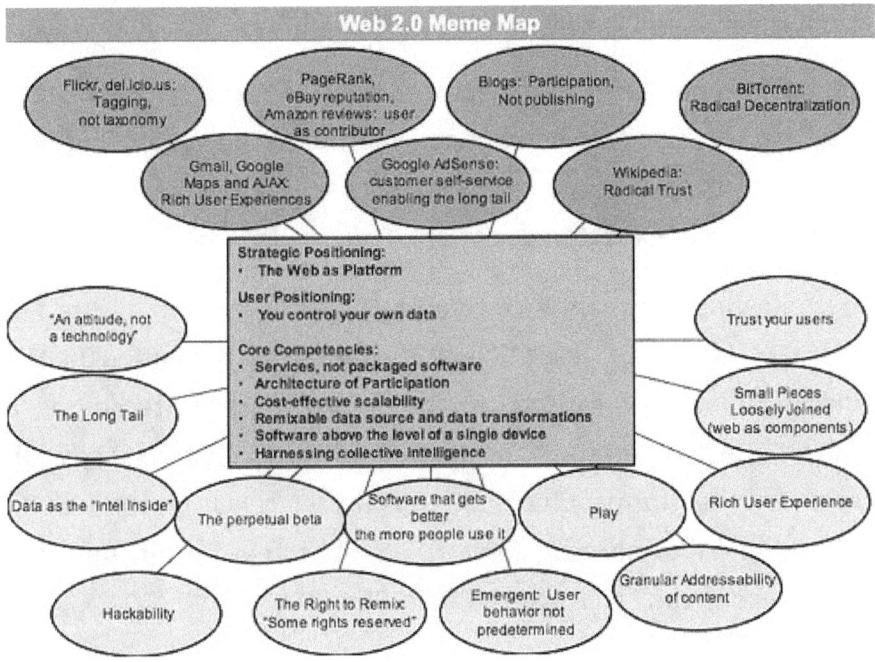

Source: O'Reilly Media Inc., 2011

Tim O'Reilly was the first who listed and described changes from Web 1.0 to Web 2.0. Furthermore a version number was utilized, which is mainly used in order to define the development of a software element at a certain point of time. At the same time the usage of a version number misleadingly suggests a new technological leap, whereas the biggest difference of Web 2.0 to Web 1.0 is its concentration on people and content rather than on defining and creating destinations for web users. (Hass, Killian, & Walsh, 2008, pp. 17-19, 74)

Though the ever fast-changing internet easily renders any emerging definition out of date, two ends of Web 2.0 definitions are from a technician's and a social scientist's point of view. The combination of techniques, architectures and technologies led to a transition to a new generation of dynamic social web applications and services. Social scientists on the other hand state that "we are the Web" and hence, whether called Web 2.0 or not, the web is shaped by people, which thus, makes it unpredictable and led in an unexpected direction. (Shuen, 2008, p. 1)

The following summary was given by O'Reilly concerning the full extent of Web 2.0: "Web 2.0 is the business revolution in the computer industry caused by the move to the internet as a platform, and an attempt to understand the rules for success on that new platform. Chief among those rules is this: Build applications that harness network effects to get better the more people use them." (O'Reilly radar, 2005-2009)

Another definition given by the internet blogger Ian Davis concentrates more on the social part of the Web 2.0: "Web 2.0 is an attitude not a technology. It's about enabling and encouraging participation through open applications and services. By open, I mean technically open, with appropriate APIs[1] but also, more important, socially open, with rights granted to use the content in new and exciting contexts." (Internet Alchemy, 2011)

In a nutshell, the term Web 2.0 cannot be defined and captured in a single, simple phrase, but is rather a combination of several

---

[1] A language and message format used by an application program to communicate with the operating system or some other control program, Source: Ziff Davis, Inc., 1996-2011.

tools and principles, more or less bound to each other. (Governor, Hinchcliffe, & Nickull, 2009, p. 1)

2.2. Web 2.0 tools

Due to the wide range of tools, which have been defined in the last years, this thesis will concentrate on those, most commonly used and described in books. The emergence of those tools, as well as private usage and business-oriented ways of how to use them for customer retention will be discussed. For further understanding one or more examples will be mentioned for each described tool.

The most well-known and most widely discussed Web 2.0 applications are Wikis and Blogs. (Damiani, Lytras, & Ordonez de Pablos, 2009, p. 171) Those two communication forms contribute to the write nature of the Web 2.0 nowadays. (Hagemann, & Voss, 2007, p. 49) Among consumers wikis and video-web pages have the highest popularity, followed by Blogs, photo-web pages and social networks. As shown in graph 1 the usage of Web 2.0 applications of companies widely differs. Podcast, only used by about 7% of internet users in Germany, but more popular in the US, is highly utilized among the "Best Global Brands 2007", with over 70%, followed by video and photo-platforms with 60%. With about 40% blogs, social communities and networks and virtual worlds and online games are listed on the places three to five. Far behind, with a utilization of 7% Wikis have the least priority for companies. (Grabenströer, 2009, p. 24)

Graph 1: Utilization of Web 2.0 applications by the "Best Global Brands 2007"

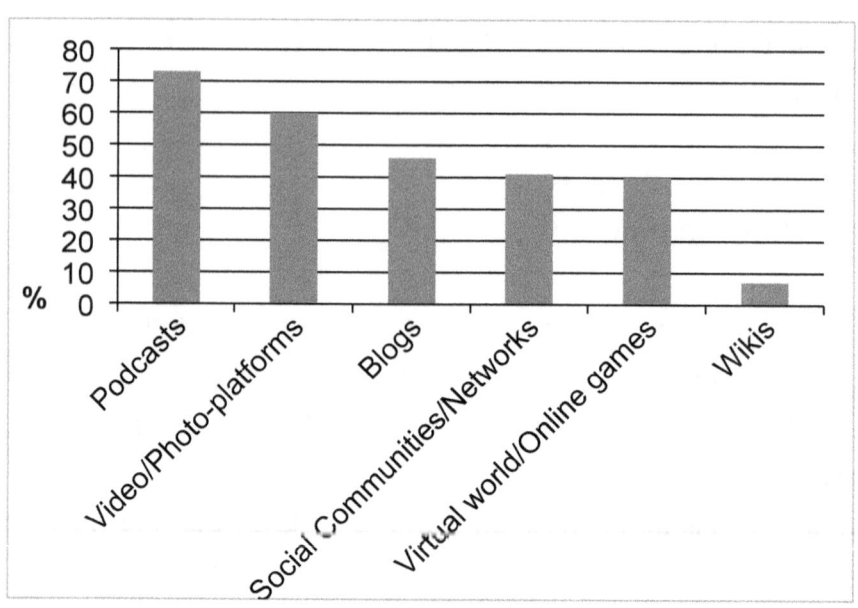

Source: adapted from: Grabenströer, 2009, p. 24

Among those companies, which are using Web 2.0 applications, there are huge differences in the branch they are operating in. The main users are computer, internet and electronic, as well as media companies, which suggests itself, since those applications offer a new channel of distribution. Due to a young and dynamic image of sporting goods' retailers, their utilization of the abovementioned tools is higher-than-average as well. Substandard users of Web 2.0 are industries such as tobacco and alcohol, which is in consequence of strict promotional restrictions. Hence not only in most of the industries, but also in the utilization of popular Web 2.0 applications as such, there is a huge potential for the future. (Grabenströer, 2009, p. 26)

2.2.1. Wiki

A Wiki is a page or a system of several pages, managed by an author in order to collect or conduct contents, which can be directly edited, and thus extended, updated, modified or deleted by other people, if it is a public one, this can even be done without registration. With this application users can contribute to and participate on the web today. (Hagemann, & Voss, 2007, p. 54) Through a wiki the provided information is either created by the user himself or in a user-driven way. Thus, additional value is offered to customers in the existing product line. (Bauer, Große-Leege, & Rösger, 2007, p. 209)

The word "wiki" originates from the Hawaiian word "wikiwiki", which means "fast" and first appeared in 1995 under the compound term "WikiWikiWeb". A distinction can be made between a sole "wiki page" and "the wiki", which is connected through several links and is a "simple, easy-to-use, and user-maintained database for creating content." (ibid., p. 54)

Appropriate wiki software allows the user to enter or edit something without using HTML, instead the modified or new content is automatically converted into HTML. (ibid., p. 54) "Wikis should not entrap to passive consummation of information like the conventional Web, but should motivate active exchange of information." (Back, Gronau, & Tochtermann, 2008, p. 10) Therefore wikis need to be designed as open systems, meaning that each and every user is able to observe and edit all the content. For special content such as headlines and emphasized words special rules apply. (Hagemann, & Voss, 2007, p. 54) Those rules are in addition to basic rules of conduct, laid down e.g. in the "Wikiquette", which is used by Wikipedia and should encourage users to treat others with respect in order to effectively build up an international online encyclopedia. (Back, et al., 2008, p. 11; Wikipedia, 2011)

Example

The most popular example for a Wiki is the abovementioned online encyclopedia Wikipedia, which was established in 2001 by Larry Singer and Jimmy Wales. Initially the Wikipedia-project should have amended the already existing Nupedia. Users should write their own contents for an international, free of charge and web-based encyclopedia. By now Nupedia lost its importance and Wikpedia, with versions in over 200 languages and more than 50.000 articles, is the biggest encyclopedia in the world. Compared to any printed encyclopedia version, Wikipedia is always up-to-date and constantly filled with news. (Alby, 2007, p. 88)

Since the quality of Wikipedia's contents is often criticized and some critics refer to it as the biggest "piece of scrap paper" in the world, a comparison between 42 articles from different fields from Wikipedia and the Encyclopedia Britannica was conducted by the journal "Nature". The result was that the number of profound mistakes was in both cases the same. (Back, et al., 2008, p. 10)

Wikis can also be used for educational purposes since they foster contribution, monitoring, communication and interaction. Wikispaces.com gives wikis free of charge and ads to educators.

Additionally there exist several free wiki tools such as: WikiMedia, TikiWiki or DokuWiki. (SiteGround.com, Inc., 2011)

2.2.2. Blog

The word Blog derives from "Weblog", resembling a nautical log written by a captain in order to document daily activities and give a description of the journey of the ship. Opinions differ about who created the term, but first blogs can be found in 1997, when people started to write about themselves and felt the need to comment on issues without any particular sense. (Hagemann, & Voss, 2007, p. 49) Blogs tend to be seen as a Web 2.0 tool, though the first blogs were created before the term "Web 2.0" has been established. (Stauffer, 2008, p. 4)

If limited to its contents a blog is some kind of diary or journal published on the WWW. Entries of a blog posted by others can be modified, though the website itself is maintained by only one individual. (Damiani, et al., 2009, p. 171) Blog entries are updated on a regular basis and the most recent post always appears first on the page. (Hagemann, & Voss, 2007, p. 50; Back, et al., 2008, p. 19)

Concurrently a new activity called "blogging" was created, which is done by "Bloggers". So the person keeping and updating the online diary is the blogger, who enters regular posts, blogs. (Hagemann, & Voss, 2007, p. 50) A blog can be conducted easily, and there is no need of programming or markup language, since a blogger can create a blog according to the WYSIWYG principle. (Koch, & Richter, 2007, p. 24)

At the beginning blogs were only used in order to report private events or state some source of information, whereas today numerous topics are discussed and hence, as shown in table 2, several types of blogs were established. (Alby, 2007, p. 24)

Table 2: Topics of Blogs

Watchblogs	Critical observation of media and literature
Litblogs	Examination of literature topics
Corporate Blogs	Blogs conducted by companies
Blawgs	Examination of judicial contents
Photo-blogs	Publication of mainly photos
Politblogs	Examination of political topics
Audio and Video-blogs	Blogs with mainly audio and video contents
Consumer blogs	Blogs conducted by consumers

Source: translated and adapted from: Alby, 2007, p. 24

Especially corporate blogs are qualified to better communicate contents and information regarding the company itself and hence clients build a rising trust in the company, which furthermore has direct effects on customer loyalty. (Bauer, et al., 2007, p. 210)[2]

Today most blogs are much more complex than the basic form of text-based entries. With images, audio or videos, graphics and photos features are added to blogs. (Hagemann, & Voss, 2007, p.

---

[2] See also: Chapter 3.3.1.2. Customer Satisfaction and Value

50) Another possibility is to enable comments, which makes a site more interactive through comments and arising discussion and can furthermore create an online community. (Stauffer, 2008, p. 12) One example therefor is the Huffington Post blog, which is, according to Technorati, currently the most popular online blog. The site offers videos, tags, RSS feeds covering areas such as politics, media, business, entertainment, living, and world news. Furthermore commenting is enabled in order to foster discussions among users. (Technorati, Inc., 2011, Huffington Post, 2011)

Critics may argue that a blog is only a forum, where the participants allocate themselves to different blogs. The essential difference compared to a forum, however, is, as mentioned before, that there is only one author and hence only the author of the original entry is able to set the principal theme. (Alby, 2007, p. 122) Furthermore forums require the user to log on, whereas for blogs this is commonly not required. The essential of a blog is the content, not the commenting. (NevOn, 2004-2006)

Example

One example is Slashdot.org, which was started in 1997 by Rob Malda in order to publish "news for nerds, stuff that matters". The blog is still maintained today, and in categories like Books, Developers, Games, Hardware, Interviews, IT, Linux, Politics or Science a very lively site was created. (Hagemann, & Voss, 2007, p. 50)

A prominent corporate blog is the one kept by General Motors Vice Chairman Bob Lutz, in which he and other GM employees inform about new developments and events in the different GM companies, respond to customers' needs and share links to videos from GMtv. With a subscription feature for users and many links to sites inside and outside GM, it is no simple blog and thus, is a new strategy for GM in order to develop their external and also internal communication. (Hagemann, & Voss, 2007, p. 50)

Examples for providers of blogs are Blogger, Blogging, Yahoo! 360°[3], or LiveJournal. Users can set up blogs within a few minutes. Those services are usually for free, whereas advertising around a blog has to be accepted. Some providers, such as Typepad, host free blogs only for a limited period of time. (Hagemann, & Voss, 2007, p. 52)

2.2.3. Micro-blogging

Micro-blogging is a recent extension of blogging and social networks and hence a new form of communication. With its momentum in 2008 micro-blogging has gained high importance especially in order to express ones opinion. Instead of whole articles micro-bloggers are only able to publish small statements or text messages, which are often limited to a specific number of characters. (Österreichische Akademie der Wissenschaft, 2011) Micro-blogs are usually answers to questions such as "What are you doing?" or "What is happening?" and can be compared to text messages, due to the limited number of characters. They describe interests and attitudes which users publish in order to share them with others. Distribution channels of micro-blogs can

---

[3] 360.yahoo.com.

be instant messages, mobile phones, emails or the Web, which deliver the comments to a user community or network. (Jansen, Zhang, Sobel, & Chowdury, 2009, p. 1)

Micro-blogging applications, such as Twitter[4], allow companies to create new ways of direct interaction with customers, which in turn can lead to an increase in customer loyalty. Information, though limited to its length and amount, can be easily and quickly disseminated and thus, in the earlier stage of the causal chain of customer retention, the creation of trust is encouraged, leading to customer satisfaction.[5] (Schmidt, 2009, p. 19)

According to Dan Connolly from Blogtronix micro-blogging cannot only help to reduce email clutter and increase productivity, but in an interview with Brian Mc Donald on March 1st, 2011, he came up with statistics including top customer and employee engagement benefits of micro-blogging based on experiences of Blogtronix's customers. Companies experienced an increase of 42% in communication with their customers, 34% higher brand awareness and customer involvement such as feedback and ideas. As described in chapter 2.4.2.2. Reichheld sees a direct link between employee satisfaction and customer retention, which is also shown by Blogtronix's statistics. An increase of 30% in employee satisfaction and a higher creativity and connectedness between staff members was stated as a direct outcome of micro-blogging. As a result 31% increase in customer retention was measured. (Square Jaw Media, 2011)

---

[4] See also: Chapter 2.2.3.1. Example
[5] See also: Chapter 3.3.1.2. Customer satisfaction and Value

Example

The first and up-to-now most popular micro-blogging platform is Twitter, which was established in 2006 and, according to statistics used at Hub Spot, in March 2010 reached the 10 billion tweets (micro-blog) mark. Registered users can publish short messages or statements with a maximum of 140 characters, which are followed by other users. Nevertheless about 55% of the 4.5 million user accounts at Twitter have never been used to twitter and another 52.7% of accounts have no followers at all. Only around 5% of all users create 75% of the tweets. (Hub Spot, Inc., 2011) Twitter has experienced an era of growth starting in November 2008, referred to as the "Twitter Red Carpet Era", when celebrities started to extensively use Tweets to keep their fan-community up-to-date. (Barracuda Networks, Inc., 2003-2011)

The hardware manufacturer Dell discovered Twitter as a sales channel and is currently supplying around 500.000 followers daily with relevant Tweets for their buying decisions. Within 24 months Dell directly gained 3 million dollars through their marketing strategy on Twitter. In the US Twitter already established as a communication channel for companies, whereas in the German-speaking countries only around 5% of the companies listed on the stock exchange actively use Twitter. (Schmidt, 2009, p. 19) Twitter quickly got high attention by companies, which discovered it for obtaining direct feedback and reaching potential customers. Using Twitter for marketing issues might seem like reputation management, but it can be much more. Users can easily access the latest news about a company, and, if used in a creative way, marketing contents are retwittered and will hence reach an immense crowd of potential customers by means of brand image. Quick and friendly customer service via Twitter can also lead to a direct increase in customer satisfaction. Special applications on Twitter can be used in order to automatically follow a targeted group or to segment your customers via Twitter lists. Another way to use Twitter is for company internal communication which can hence increase employee satisfaction and loyalty. (Weinberg, 2010, pp. 159-161)

More open source micro-blogging applications are: (WebResourcesDepot, 2011)

- Status.net offers a smooth usage and is a widely used, mature application.
- Yonkly, besides the open source version, also offers a hosted and an advanced standalone version.
- Jaiku gives anyone the possibility to set up and run their own **JaikuEngine** instance on Google App Engine and the usage of open source mobile client and frontend.
- Floopo has a function similar to Twitter and provides a powerful backend to manage the whole system.
- Blurt.it is a fresh micro-blogging application requiring PHP/MySQL to run and offers similar features to Twitter with open and private threads.

2.2.4. Community and Social Network

An online or virtual community exists when a group of people, who share a topic for a certain time on a voluntary basis, regularly meet on a computer-based way and so, due to their interactions and online activities, develop a personal relationship and an adequate feeling of community. (Back, et al., 2008, p. 66)

Already in the 1960s J.C.R. Licklider had the vision that people will use networked computers not only in order to get information, but will be able to communicate with people in different places. With the invention of the Arpanet, the predecessor of the internet, in 1970 this vision became true and since then it rapidly developed. (ibid., p. 66)

Social networks can be referred to as communities of practice, which "are groups of people who share a concern or passion for something they do and learn how to do it better as they interact regularly." (Learning Theories, 2008) According to Wenger, 2008, three crucial characteristics have to be fulfilled by an online community: (Learning Theories, 2008)

- The domain: There is one domain, which has to be shared by all the members as a common sphere of interest and through which the community is identified.
- The community: It consists of members interacting, sharing information, discussing about the domain of interest in order to satisfy their needs, take over particular roles or to learn from each other.
- The practice: Participants, members, develop a shared directory of resources concerning their practices

(experiences, documents, tools, etc.) and set certain rules concerning the communication and interaction within the community.

Furthermore a computer system is needed in order to support the community and the social interactions and to foster the community spirit. (Back, et al., 2008, p. 68) A community is not defined by the use of Web 2.0 technologies, but they facilitate the creation of such communities. (Damiani, et al., 2008, p. 171)

When looking at companies' communities it can be distinguished between active and passive ones: (Bauer, et al., 2007, pp. 205-208)

 − Active Communities: the provider of the community is actively participating in it and publishes contents in order to influence other users.
 − Passive communities: the provider does not participate himself, but only monitors the therein published contents and activities.

The active participation and interaction between the user community and the published content can foster customer loyalty. But also the monitoring of the interaction among users contains useful information about certain opinions and preferences. The successful conversion of this information into the assortment and pricing policy can lead to an increase of a user's trust and hence also his acceptance towards minor errors. Furthermore, in the stage of trust and acceptance, the perception of a user toward the company can be influenced via social networks, which is an important issue from a customer's point of view. (Bauer, et al., 2007, p. 209)

Example

As the biggest social network worldwide Facebook is one of the most popular examples. Founded in 2004 by Mark Zuckerberg, a Harvard student in his studies, its user number quickly increased and was extended to other universities and schools in the US. Today Facebook counts over 400 million active users, of which 70% live outside the US and it is growing continuously. (Digital Buzz Blog, 2008) As a major step to Facebook's success the platform was opened for developers to allow them to build custom applications. Furthermore a new feature was added to social networks by Facebook, where users were able to develop a network of existing and new friends. (Newman, Thomas, 2009, p. 42)

In order to properly use Facebook for marketing issues, companies can either create a page, a group or use paid ads. Pages, similar to profiles, but for businesses, organizations and public figures, can be liked by anyone, are free and can hence be set up easily. Nevertheless it is hard to build a fan-base via pages and to stick out from the huge amount of pages on Facebook. If information is provided useful, informal and in a friendly way, pages can act as interaction platforms for e.g. contests, which can actively foster customer loyalty. Apart from that results and statistics provided by Facebook can be used to gather information about your customers. Nevertheless one has to find the right balance between providing information and spamming customers. Groups, on the other hand, are similar to discussion forums with additionally a wall in order to offer fans of a group the possibility to give their opinion and interact with them. Free of charge groups have also a high level of engagement including discussions and active participation, but can be very time consuming. For targeted marketing Facebook ads can be used in order to reach people from the same geographic region, groups or even with the same educational background. Though ads are powerful targeting parameters, they can get very expensive, depending on a company's goals. (KISSmetrics, 2011)

The most popular social networks ranked after Facebook are: Twitter[6], MySpace, LinkedIn, and Ning. Facebook itself offers the possibility to connect with friends all over the world and due to its huge user base people come back to the side. Twitter offers most recent updates on events and also on stars' life, whereas MySpace main advantage over other social networks is the possibility to customize and design one's own page. LinkedIn is primarily used for creating business contacts and broaden users' professional interests online. The network Ning was popular for its free sites, when in April 2010, a new CEO decided to charge users and thus, now focuses more on delivering features and services which benefit and help users and are hence positive for the premium service customers. (Ning, Inc., 2010; eBizMBA Inc., 2011; Yahoo! Inc., 2011; Boutell.com, 1994-2011; WebBizIdeas.com, 2011; Nick Tadd, 2011)

More than any other Web 2.0 tool communities and social networks have an emotional selling proposition, meaning that

---

[6] See also: Chapter 2.2.3.1. Example

users feel part of a unique group, hip, trendy, remembered and accepted. (Bulk Email marketing Tips & Services, 2009)

2.2.5. Podcast

Podcasts can be seen as small radio or TV shows which can be consumed online independently to airtime. (Back, et al., 2008, p. 51) Most podcasts are free of charge and can be created and published by everyone about any preferred topic. The term podcast combines the name of the MP3-player "iPod" and "Broadcasting". (Alby, 2007, p. 73) In 2004 Adam Curry and Dave Winer designed the software which built the basis for podcasting as we know it today. (Back, et al., 2008, p. 51)

It is important to mention that most video-podcasts are accessed through Feeds, such as RSS, a family of formats to present data, since otherwise the users could not subscribe and the podcast itself is not listed elsewhere. Whereas audio-podcasts can be downloaded without using any Feed. A podcast itself indicates an episode of video or audio files. (Alby, 2007, p. 73)

Podcasts have the advantage that users can choose which contents they want to watch or subscribe to. The thereof obtained information can be used by companies in order to infer to customers' preferences. Furthermore podcasts can act as a channel to push information transfer, which in turn can increase customer satisfaction, due to supporting information and thus, can lead to guidance of attitude. (Bauer, et al., 2007, p. 209)

Example

34

The search engine Podscope.com offers the possibility to search for certain contents. Remarkable is, that the podcasts are searched for spoken words used in the video or audio clip. An example therefore is the podcast of the Austrian radio station Ö3.

Some other podcasts, which were recently developed on the Austrian platform podcast.at are: stereopic, Wort-Laut, Eine Zigarette mit Karsten – Der Podcast, or Hagazussa-TV. (podcast.at, 2010)

2.2.6. Video and Photo Platforms

The first online video and photo platforms as commercial software products were developed in the early 2000s when the video and photo usage via internet massively expanded. Such platforms host videos or photos of registered users, who are either private individuals or organizations. Video platforms can be self-serviced, which means that a standard video player is used in order to facilitate the creation, editing, tracking or delivery of a video. Another form is the serviced platform, which is mainly offered by companies in order to be able to customize and take part in customers' video platforms and hence foster the required workflow. Users of video and photo platforms can leave comments on the content published on the website. Many social networks also offer the possibility of uploading photos and videos, and are thus, no pure photo or video platforms. (Flimp Media, 2011; Der Standard, April 2010)

Video and photo portals, similar to podcasts and newsfeeds, offer the possibility to supply customers with additional information and thus, can foster customer satisfaction and steer their attitude towards the company. (Bauer, et al., 2007, p. 209)

Example

The video platform YouTube was launched in 2005 by former PayPal employees and in October 2006, though till then only deficits had been produced, was bought by Google. YouTube is used by private users and companies, who upload their home-videos, music and commercial videos, as well as film and TV clips to the platform. As the biggest and most popular video platform every day around 2 million videos are being watched and about 100.000 new ones are uploaded to YouTube. Though its big success, YouTube already deleted 30.000 films and video clips, which infringed copyrights in 2006 and is continuously being charged for copyright infringement. (Sauer, 2006, p. 65; Völtz, 2011, p. 1)

Other examples for video platforms are: Sevenload.com, Yahoo Video, AOL Videos, MSN Videos, Google Video, Videos on MySpaceTV.com or videotube.de. (www.IHans.de, 2003-2008)

The Web-based community Flickr was launched in 2004 by Ludicorp in order to organize and share photos and already in 2006 the 100-millionth photo was uploaded to the website. The photos on Flickr remain the property of a user, but can be published under certain licenses, with descriptions and identifying tags added. (Hagemann, & Voss, 2007, p. 186)

Some other photo platforms are: Photofolio, Sharpfolio, Tumblr, or Pixel Post. (MetaFilter Network, Inc., 1999-2011)

2.2.7. Virtual World and Online Games

Online games are seen as the new emerge for viral marketing since the first online games launched in 1995. However online gaming already started back in 1969 when a two-player version of the game Spacewear was established. Nowadays online games are most of the times free and heavily used among the male 16-25 generation. More and more companies also take advantage of marketing via online games, which increases the mean retention time on corporal websites and can lead users successfully through the online buying process. A good online game which entertains users is likely to be recommended and customers tend to reveal their personal data and other relevant marketing data more easily. Furthermore interaction, dialog and involvement with the brand are relatively inexpensive and cost effectively fostered. Equally as important is the association of the brand with fun, excitement, entertainment and sometimes even learning. On social network sites such as Facebook, online games can be easily promoted and spread and can support the already existing marketing on the respective site. But online games can also be used in the recruitment process and the on-job training for employees, which according to Reichheld's Loyalty Effect[7] has also direct impact on customer retention. (Illustree, 2007;

---

[7] See also: Chapter 2.4.2.2. The Loyalty Effect.

EzineArticles.com, 2011; Alf Nucifora, 1997-2009; Harvard Business School Publishing, 2011)

Example

Chrysler started a successful online game in order to increase the brand awareness among middle-aged women. "Get Up and Go" was following the popular "Cosmo Quiz" format by matching each player to a travel personality and a corresponding Chrysler vehicle. The game was e-mailed to retailer stores which put links on their Web sites. The time spent on this online game was on average around 7 minutes and 15 times more players than overall Web sites visitors requested vehicle brochures. Players were furthermore encouraged to invite friends in order to compare and match results. Chrysler also collected the players' email addresses and some demographic information and could at the same time display information about the company's products. (Harvard Business School Publishing, 2011)

Chrysler furthermore utilizes online games in order to train Jeep and Dodge dealers on different four-wheel-drive systems, which resulted in a ten times higher retention rate from playing than from reading a mail. (ibid.)

Some other examples for corporate online games are: "Kingdom of Gondal", which was published in February, 2011, by the Bob Mobile AG, "Webspecial 15 Jahre Max Maulwurf" by Deutsche Bahn AG, "100 Years Synthetic Rubber" by Lanxess Deutschland GmbH, "Mountainbike-Excursion" by Florena Cosmetics GmbH or "Schießbude" by Freiberger Brauhaus GmbH. (Bob Mobile AG, 2011; Internet und Werbeagentur Dresden, 1999-2010)

Second Life is a free 3D virtual world where users can socialize, connect and create using free voice and text chat, which was launched in 2003 by Linden Lab. Free client programs, called Viewers, are offered, which enable Second Life users, Residents, to interact with other users through avatars. (Linden Research Inc., 2011)

2.2.8. Newsfeed

A newsfeed is a data format, which is used to spread frequently changing contents of websites among interested users. Several contents are bundled up to a feed, which then can be subscribed by users. Users will be automatically informed about any new contents published at the website with no need to recall it. Feeds can also include multimedia-contents, such as audio or video files, which subscribed users can download by means of respective programs. Furthermore newsfeeds are based on the pull-principle, meaning that subscribed users really just get the content they subscribed for. In case the user is not interested anymore in a feed the subscription can be cancelled easily. (Back, et al., 2008, p. 57)

Like podcasts, newsfeeds offer the possibility to gain information about users' preferences and companies are able to feed the community with appropriate information in order to increase customer satisfaction and influence their perception about the company itself.

In order to properly understand a newsfeed its lifecycle needs to be considered:

1. Publication: The newsfeed is created and published so that a broad user community is reached as fast as possible. Due to this, short fugacity publications such as current price lists or special discounts are more applicable for newsfeeds than others. (Back, et al., 2008, p. 59)

2. Brokering: Since a huge amount of newsfeeds exist on every website, it is not guaranteed that contents are actually read by users, which is also due to the abovementioned pull-principle. Brokers take over the role of distributors. They identify the source, support the cooperation between client and server, arrange the registrations and spreading of contents and delete inoperative services and feeds. The goal is to win as many clients for one's own feed as possible. (ibid., p. 59)

3. Aggregation: In the last phase the created newsfeeds is consumed and their value is dependent on the adequate contents according to the recipient's needs. News-aggregation is mainly done by knowledge workers, who research for external information relevant for their company. (ibid., p. 59)

Example

Newsfeeds are widely-spread today and cover all different kind of topics. The New York Times newsfeed covers topics such as: news, culture and lifestyle, and marketplace as shown in figure 2.

Many newspapers already offer newsfeeds, such as Kurier, Welt Online, Standard, etc. But also libraries like the Austrian Central library for Physics, or the Johannes Keppler Universität Linz.

Figure 2: New York Times Newsfeeds

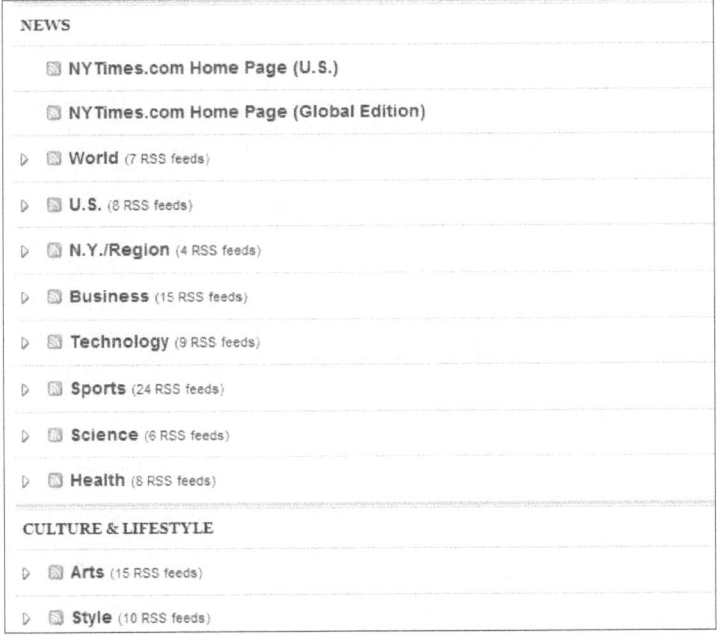

Source: New York Times Company, 2011

2.2.9. Tagging

Tagging is the act of a community, built of a system's users who characterize the objects of this system with metadata such as keywords and categorizations. Typical objects are pictures, audio and video files, bookmarks, Weblogs, recommendations for literature, wikis, text documents or ideas. The keywords used are called tags. (Back, et al., 2008, p. 41) Since there is always an emotional or irrational reason for adding a tag, they can only be used by human beings and not by machines. (Hagemann, & Voss, 2007, p. 191)

There are two different motivations behind tagging, and hence it can be distinguished into social and collaborative tagging. Whereas collaborative tags are an act of sharing an organization's scheme and information, the social drivers are attention, self presentation, opinion expression or just a byproduct to play and competition. (Hagemann, & Voss, 2007, p. 185)

From a business point of view tagging can be used as an additional tool in order to provide relevant information to customers, which can primarily be used in the front-end policy.[8] Furthermore the targeted audience based on their interests can be reached more easily. (Bauer, et al., 2007, p. 209)

A collaboration of all the tags used by different users for an object is called folksonomy. It is the process of making all tags on an object visible to other users in order to adopt tags available for all users of the website. The term is compounded of "folks" and

---

[8] See also: Chapter 4.1.4. Front-end policy

"taxonomy", a term which was already existent in Web 1.0 and, compared to tagging, offers a framework for a maximum readership. Two structural elements used for folksonomy are "Tagging" and "Instant Feedback", which shows all assigned tags coherent to a keyword. This has the effect of alleviation of tagging and promotes a "silent dialogue" between users. (Hagemann, & Voss, 2007, p. 183) In order to be able to visualize folksonomies, a tag cloud is used, listing the tags in a one or two dimensional alphabetic order. As shown in figure 3, the more a certain tag has been used, the bigger or more noticeable it will appear within the tag cloud. (Back, et al., 2008, p. 41)

Figure 3: All time most poplular tags

animals architecture art asia australia autumn baby band barcelona beach berlin bike bird birds birthday black blackandwhite blue bw california canada canon car cat chicago china christmas church city clouds club color concert dance day de dog england europe fall family fashion festival film florida flower flowers food football france friends fun garden geotagged germany girl girls graffiti green halloween hawaii holiday house india iphone island italia italy japan kids la lake landscape light live london love macro me mexico model mountain mountains museum music nature new newyork newyorkcity night nikon nyc ocean old paris park party people photo photography photos portrait raw red river rock san sanfrancisco scotland sea seattle show sky snow spain spring square street summer sun sunset taiwan texas thailand tokyo toronto tour travel tree trees trip uk urban usa vacation vintage washington water wedding white winter woman yellow zoo

Source: Flickr, 2011

Example

As discussed in chapter 2.2.6.1. the Web-based community Flickr also offers the possibility of tagging photos which are published on the website. The published contents are hence categorized through user-defined keywords, which can be chosen without any restrictions of choice but only permitted to the author himself. (Hagemann, & Voss, 2007, p. 186)

Other examples for applications using tagging are: del.icio.us[9], blogg.de, or taggling.de. (Internetberatung Sven Przepiorka, 1998-2011)

2.2.10. Mash-up

First used in the music branch the term "mash-up" meant the illegal creation of a new song by using two existing ones. In the context of Web 2.0 mash-ups are described as "web applications that merge data from one or more sources and present them in a new way", which means the creation of new content. By providing APIs, the owners of the data can facilitate and encourage the usage to other users and hence minimize the expenses to create a new offer. Mash-ups have just developed in recent years and individuals, combining existing data in untraditional ways, come up with new innovations every day. (Chow, 2007, p. 7)

By combining special applications, mash-ups can be supportive for conducting transactions. Benefit is added by placing relevant

---

[9] See also: Chapter 2.2.11.1. Example

information about buying processes at users' disposal. In special cases mash-ups can foster interaction and thus, can raise acceptance, what in turn, can have a positive influence on customer loyalty. (Bauer, et al., 2007, p. 210)

Example

The launching of the new Nintendo Wii in November 2006 preprogrammed a shortage in supply. With the Wii Seeker site the search for Wiis should be facilitated through the combination of expected shipment information, target stores and Google Maps. Markers on Google Maps represented the target stores and offered information about it such as the address and the number of Wiis it was expected to have on the particular day chosen. Furthermore stores close to the potential buyer's location were listed and trips there could be planned in order to maximize the chance of actually getting a Wii. After the launch a reinvention took place by adding auction information from eBay and product information from Amazon. Customers can now post notes to one another about the stores' inventories. (Chow, 2007, p. 7)

Other examples for mash-ups are: (Chow, 2007, p. 12)

- Astrolicio.us combines data feeds from sites like Digg.com, GoogleNews or Videos and presents astronomy news for vistors.
- Popurls.com collects URLs from popular sites.
- Housingmaps.com plots housing listings from Craiglists onto a map.

- Keegy (us.keegy.com) aggregates news and personalizes it for users.

- Alkemis (local.alkemis.com) aggregates and maps e.g. pictures and life web cams in selected cities.

- Gametripping.com offers a collection of satellite and Flickr photos of baseball stadiums.

-

2.2.11. Social Bookmarking

Social bookmarking can be referred to as bookmarks, which are formed through collaborative indexing of a net's users. Bookmarking systems such as del.icio.us manage, exchange and save electronic bookmarks and furthermore offer the possibility to save references on websites together with keywords, called tags. So users, tags and resources, like bookmarks, are the basic elements of bookmark-systems connected to each other. When a tag is assigned to references through a user a folksonomy is created. (Back, et al., 2008, pp. 26-29)

Example

The website del.icio.us was launched in 2003 in order to manage large amounts of collected bookmarks, which users can browse through. Usually free for everyone, for access to some bookmarks users need to get registered. The social bookmark system of del.icio.us can be divided into the basic elements bookmarks. e.g. web addresses of web sites which users consider valuable, and metadata, e.g. data on the bookmark data. (Hagemann, & Voss, 2007, p. 196) The basic idea of del.icio.us is the abovementioned

Folksonomy-principle, which allows the tagging of bookmarks. (Back, et al., 2008, p. 31) Though the former success of del.icio.us Yahoo! now decided to shut it down and many people try to find alternatives for where to move their bookmarks to without losing their tags. The closure of del.icio.us can be seen as the end of the glory era of bookmarking services, which was also marked by the improvement of social networks on sharing links. Del.icio.us has not been the first website owned by Yahoo! to be laid off and will also not be the last. Users of Flickr, which is also owned by Yahoo!, already seek an alternative website, just in case Yahoo! decides also to shut this one down. (AOL Inc., 2011)

Some important functions of del.icio.us were: global tag clouds and the ones related to users[10], separation of tags through bookmarks, or sending of links with special tags to other users. (Back, et al., 2008, p. 31) Del.icio.us was also utilized to find out more about customers' interest and preferences and through its tags it was useful in order to reach the targeted clients according to their interests with valuable information. (SlideShare Inc., 2011)

Other social bookmarking applications are Mister Wong, which is the European pendant to del.icio.us, or Digg.com, where links videos, podcasts and blogspots can be inserted, evaluated, commented on and also followed by subscribing to other users' channels. (Mister Wong, 2011; Digg.com, 2011)

---

[10] See also: Chapter 2.2.6. Videos and Photo Platforms

2.3. Summary

The previous chapter was concerned with some of the various definitions of Web 2.0, stating the most comprehensive ones, with a focus on the term's history. Special emphasis was thereupon put on Tim O'Reilly's demarcation of Web 2.0, who, like other specialists, came to the conclusion that no single and commonly effective definition can be given.

Furthermore the most commonly defined Web 2.0 tools were described in detail, such as Wiki, Blog, Social community and networks, Podcast, Newsfeed, Tagging, Mash-up and Social Bookmarking, by means of practical examples and their field of application in the customer retention processes.

Hence, the following research questions were answered in this chapter:

- What is Web 2.0?
- Which Web 2.0 tools are known and how are they defined?

3. Customer Retention

Defining the term "customer retention" is necessary for the onward approach of this paper – especially for the establishment of the outcome of this thesis in the last chapter – but also in order to acquire a basic understanding of the stated problem.

To sufficiently cover the term and its surrounding field and also as an essential for the further approach, special attention is paid to managing customer retention.

3.1. Definition

The term customer retention is in literature often misleadingly equated with expressions such as customer loyalty, relationship management, retention marketing, brand and product loyalty as well as customer satisfaction. (Bruhn, & Homburg, 2005, p. 8) Though not exactly the same, there appears to be an alliance between loyalty, satisfaction and retention. (Allen, & Rao, 2000, p. 7) Bliemel and Eggert (1998) see customer retention as a "system of activities for enhancing the transaction process, on the basis of the positive positioning of the customer, and the resultant's readiness for successive purchasing" (Bliemel, & Eggert, 1998, p. 38). With more emphasis on the guidance of customers, according to Diller, 1996, as well as Meyer, Oevermann (1995) customer retention, also referred to as customer commitment management, comprises: "all of a company's measures which are targeted at the positive shaping of the customer's current behavior patterns, as well as the future intentions of the customer with regard to the provider, or its services, in order to stabilize or expand future relationship with the customer". (Bruhn, & Homburg, 2005, p. 8) These repurchase activities and behaviors can be seen as an outcome of loyalty. (Allen, & Rao, 2000, p. 8)

Two main perceptions related to the term customer retention are the measurement-oriented and the behavioral perspective: (Werani, 2003, p. 1)

From the measurement-oriented perspective a bunch of suppliers' activities should lead to a tight relationship to customers and hence obviate a loss to competitors. Activities taken can on the one hand lead to loyalty as well as to constraint, the latter of which ads a negative connotation to retaining customers. (Dittrich, 2002, p. 2) From a customer's perspective emanating activities can lead to retention in two different modalities, namely either being voluntary dedicated or being involuntarily bound to a company with high costs and difficulties involved in change or no alternatives available. A constraint is most likely in oligarchic political systems or monopolies. Within loyalty one further has to discern between rational, cognitive and emotional, affective loyalty, whereas the latter has to be given the highest importance. Customers will only be loyal, if their satisfaction is so high that there is no wish to change. (Ajami, Gargeya, Goddard, & Raab, 2008, p. 80)

The second approach is the behavioral perspective, which concentrates on customers' attitude towards a repurchase. Thus customer retention is seen as the effect on customers, which has to be achieved through measurements from the supplier's side in order to foster customers' involvement, commitment, trust and satisfaction. (Werani, 2003, p. 2, Diller, 1996, p. 88)

3.2. Opportunities and Threats of customer retention

Customer retention constitutes the following opportunities and threats as defined in table 3 as well as discussed in detail thereinafter:

Table 3: Opportunities and threats of customer retention

Opportunities	Threats
1. Effective customer information	1. Loss of information and flexibility
2. Potential cost reduction	2. Loss-generating customer retention and unremunerative customers
3. Higher intention of recommendation and repurchase	3. One-sided customer structure
4. Cross selling – potentials	4. Omitted and aggravated customer acquisition
5. Reduction of overall business risk	5. Neglected customers
6. Customers involved in co-determination	6. Resistance of customers
7. Loyal employees	7. Ineffective commitment programs, Zero-sum game
8. "Quasi-monopoly"	

Source: translated and adapted from: Kindermann, 2006, p. 9

3.2.1. Opportunities:

1. Effective customer information is offered by airing one's products and services to customers and, thus, no further information about the company as well as about competitors are gathered. Furthermore buyers are more willing to share information with suppliers and hence suppliers can respond accurately to customers' needs and expectations and alleviate the risk of focusing on the wrong production. (Dittrich, 2002, p. 15)

2. Long-ranging customer retention concurrently cuts overall costs for the acquisition of new customers and thus, constitutes potential cost saving. The longer a customer is retained to a company, the less costs for customer service incur, and a 5% increase in customer retention can result in over 25% of increase in profit. (Reichheld, 1996, p. 8) Furthermore caring for loyal customers is about 15-20% less expensive than new customer acquisition. (Finkelman, & Goland, 1990, pp. 2-4; Müller, & Riesenbeck, 1991, p. 69)

3. Compared to neutral customers, satisfied customers tend to refer the company and also repurchase more often, which results in automatic proliferation of information about one's products' features and hence entails new potential clients. (Dittrich, 2002, p. 15)

4. "Customers' profitability rates tend to increase over the life of a retained customer." (Business E-Coaching, 2010) Hence, additional market potential of customers can be gained through long-term customer retention by selling

more frequently, in bigger amounts or different products to one client.

5. The more positive a customer's attitude towards one's products and services is, the more tolerant he becomes to errors or unsatisfactory service from the supplier's side. Greater reciprocal trust is built through numerous interactions entailing long-term relationships. (Ajami, et al., 2008, p.82)

6. Buyers actively contribute to co-determination of the supplier-customer relationship and can even propose ideas for improvement when they feel bound to their supplier. (Dittrich, 2002, p. 15)

7. By succeeding in customer retention and hence supplying satisfied customers also employees become more loyal to the company and work more motivated and enthusiastic. (Dittrich, 2002, p. 15)

8. A "Quasi-Monopoly" is generated, when customers do not desire anymore to conduct extensive research before buying, but even blend out competitor's products. At this stage, the customer retention is of a height where one company is perceived as the only option. (ibid., p. 15)

3.2.2. Threats:

1. In case of a very strong buyer-seller-relationship in B2B business, the risk of employees' defection is given, which would accordingly lead to a loss of information and know-how. (Dittrich, 2007, p. 17)
2. A very strong focus on a specific group of customers can lead to a loss of information and flexibility through late or no recognition of recent developments. (ibid., p. 17)
3. A one-sided customer structure could result from positive word-of-mouth by regular customers of a specific age class by only recommending to people of the same age. (ibid., p. 17)
4. Sole concentration on customer retention has the effect of neglected customer acquisition and in the long run will hamper future relationships with new customers. (ibid., p. 18)
5. Concentrating all efforts on several main key-accounts might disregard other customer groups with future potential and can result in a loss of the aforesaid. This potential threat especially exists in the case of differentiated actions for customer retention. (ibid., p. 18)
6. As mentioned before, customer retention can be perceived from two sides: namely either on a voluntary basis, or as being constraint. The latter might be experienced when a customer's perception is not adhered, and hence would result in customer resistance. (ibid., p. 19)

7. Some suppliers buy concessions of companies in order to reach customer retention to full capacity. Investing into customer-supplier relationship at any cost can be ineffective and result in a zero sum game, or even a loss. (ibid., p. 19)

3.3. Effectiveness of customer retention

In order to give a broader view and to offer conceptualization of the effectiveness and impacts of customer retention, the causal chain will be defined by means of two comprised theories. Those theories of causal chains of customer retention descend from Allen and Rao, 2000, and Möhlenbruch, Dölling and Ritschel, 2007, and should replenish each other as for defining the process.

As a basis for this thesis the graphic of the causal chain from Allen and Rao will be taken and determinants and focus in the respective stage of customer retention will be replenished by Möhlenbruch, Dölling and Ritschel's concept. Both theories are divided in a four-stage approach to customer retention. Whereas Allen and Rao focus on the initial contact, Möhlenbruch, et al. mainly concentrate on the satisfaction and loyalty stages and respectively add important determinants. Hence, combining the two approaches gives a well-balanced composition of essential determinants for customer retention.

3.3.1. Causal chain of customer retention

The "Causal Chain of Customer Retention", given in figure 4, shows the correlation of four critical predictor variables and two sets of intermediate variables, which in the end result in customer retention. (Allen, & Rao, 2000, p. 9) Additionally the focus and the various measurements taken at each single stage will be defined from a supplier's point of view. (Dittrich, 2002, p. 19)

Figure 4: Causal Chain of Customer Retention

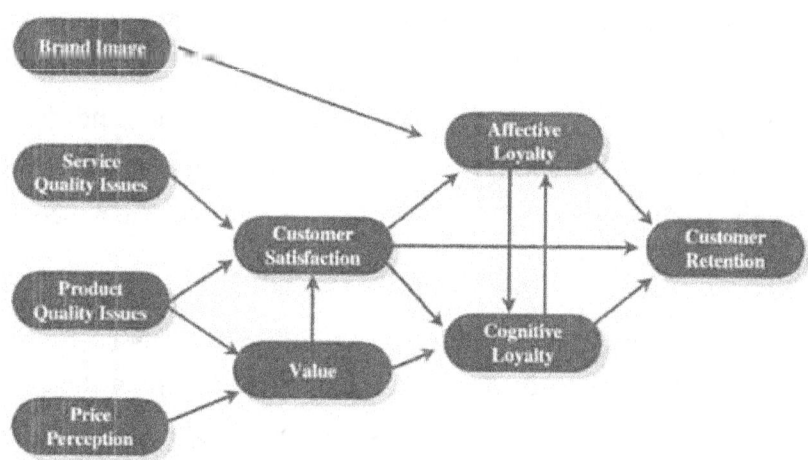

Source: Allen, & Rao, 2000, p. 9

Basic determinants of customer retention

Three of the four critical predictor variables, which are service quality issues, product quality issues and price perception, determine the basis for the creation of customer satisfaction and value. The fourth variable, brand image, directly impacts loyalty, but solely on the emotional side. (Allen, & Rao, 2000, p. 9) At this stage of the causal chain an initial contact between a product or a company and their customers is made. (Bauer, et al., 2007, p. 200)

Customer satisfaction and Value

The first stage of action concentrates on customer satisfaction as an objective. In order to achieve satisfaction comparison between actual and target state of a product has to be drawn. (Bauer, et al., 2007, p. 200) Service and product quality issues, which can be subsumed as actual brand performance, have to exceed a customer's perceived brand performance in order to satisfy customers. The "Confirmation - Disconfirmation Paradigm" shows that by only meeting the expected brand performance, customers tend to have neutral, indifferent feelings about a product. (Customer experience labs, 2008) In most cases customer satisfaction is perceived as necessary but not sufficient for developing loyalty. Hence value needs to be added through product quality and price advantages. The perceived value of a product on the one hand influences the satisfaction stage itself, but also leads to rational loyalty. (Allen, & Rao. 2000, p. 10) Particularly important in this stage is information provided by the company, so that clients can build the right expectations concerning the brand performance. (Bauer, et al., 2007, p. 200)

The loyalty stage

Though often equated, a clear distinction between satisfaction and loyalty can be made. Being geared specifically to product and service attributes, customer satisfaction is perceived as a relatively dynamic measure. Loyalty, on the other hand, describes a wider range of static attitudes towards a company in general and differs from satisfaction in terms of trust, acceptance and the formation of enthusiasm of the customer. (Bauer, et al., 2007, p. 201) Furthermore loyalty can be distinguished into emotional and rational components[11] and is clearly influenced by satisfaction. Customer satisfaction, perceived value as well as brand image directly impact loyalty. Most probably a mutual causation between affective and cognitive loyalty exists, which would mean that they procure one another. (Allen, & Rao, 2000, p. 10)

Trust is mainly built through reliable information and events from the supplier's side, which positively affects voluntary commitment of customer. Acceptance on the other hand defines a client's willingness to disregard minor errors and competitors' hostilities. In this phase the involvement as well as advancement of engagements, thus the interaction between customers and suppliers, is of utmost importance. (Bauer, et al., 2007, p. 201)

The final outcome: customer retention

The final outcome, customer retention, is directly impacted by customer satisfaction and loyalty. In most industries customer retention is measured attitudinal, since a well-established customer database capturing also actual customer loss is required, in order to behaviorally measure the rate and its tangents, like share of wallet,. (Allen, & Rao, 2000, p. 10) Customer retention is seen as executed when repurchase, cross-buying, recommendation and reduced price-sensitivity are performed by customer. Thus transaction is the main focus of this stage. (Bauer, et al., 2007, p. 202)

---

[11] See also: Chapter 3.1. Definition

3.4. Managing customer retention

Using customer retention as a strategy means to segment and differentially target a company's current customers in order to increase the value of a relationship for both seller and buyer. (Baran, Zerres, C., & Zerres, M., 2009, p. 3) Systems in order to support relationship-based customer service require strategic planning. Maintenance of the same, on the contrary, requires "constant focus on indicators of the quality of customer management" which are mainly market-research-based measures of loyalty, customer-database measures of retention, and system-based measures of complaints and their resolution. Hence managing customer retention involves all actions taken by the suppliers in order to successfully retain customers. (Bruhn, & Homburg, 2005, p. 17)

3.4.1. Fundamentals of managing customer retention

A utility should not aim to retain all customers, but focus on those, most likely to respond to incentives, differentiated service or other methods. Hence exchange of information is fundamental for retention in order to bridge a customer's state of mind and his actual behavior. If properly managed, a supplier should be able to influence customers over the long run to repurchase or at least sustain their current business level. Nevertheless, without a solid customer relationship foundation, investments into higher customer service levels would be a waste of resources. (Foss, & Stone, 2001, pp. 484 – 485)

The better you know your clients and trust in them, the more predictable and efficient and hence also the more profitable doing business gets. Thus managing customer retention should be of utmost importance for every company, since, when consistently high, it can lead to tremendous competitive advantage, increase in employee morale, higher productivity and growth, as well as a reduction of capital costs. (Reichheld, 1996, p. 2) Empirical studies show that an increase of five percent in customer retention rates can lead to a value rise of an average customer by 25 to 85 percent in a wide range of industries. (Dawkins, & Reichheld, 1990, p. 41; Töpfer, 1996, p. 92) Marketing costs in order to get a customer to reorder from your company are on average five to six times lower than for new acquisitions. (Allen, & Kania, 2001, p. 7) Especially in saturated markets or markets of a shrinking number of new customers, customer retention ever increasingly is seen as a core business operation. (Ahmed, & Buttle, 2002, p. 152)

3.4.2. Different theories of managing customer retention

Since there is not one single theory about how to manage customer retention, two different approaches were chosen, which should amend one another. Whereas Bruhn and Homburg (2005) concentrate solely on retaining customers, by defining 6 dimensions, Reichheld (1996) determines customer retention to be interrelated with retention of employees and investors.

The Dimensions of customer retention

As shown in figure 5, the six different dimensions of customer retention need to be taken into consideration when determining the strategy:

Figure 5: Dimensions of customer retention strategies

Source: translated and adapted from: Bruhn, & Homburg, 2005, p. 17

The most important steps in order to set the right strategy are: (Bruhn, & Homburg, 2005, p. 18)

1. Determine the reference object of customer retention: Which subject should the customer be bound to?
2. Define the target group: priority of investments in different customer segments.
3. Identify the method of customer retention: How should the customer be bound to your company?
4. Define the instruments of customer retention: Which instruments should lead to being retained? How are those instruments designed (focused on dialogue and interaction or rather on building up switching costs)?
5. Determination of intensity and timing: e.g.: Sending out newsletters every two weeks or once a year?
6. Coordination of the own strategies: agree about the chosen strategies with your sub-supplier and retail seller in order to deploy the effect of synergy.

In order to show the different definitions and ways of how to manage customer retention the following chapter 3.4.2.2. will cover Reichheld's loyalty effect theory. Whereas alike six, , strategies for customer retention are defined, comparable to Bruhn and Homburg, Reichheld additionally lists measurements of how to attract and retain high-quality employees. Seeing the retention of employees as an essential for customer retention, Reichheld shifts its management into the Human Resource department's scope of functions.

The Loyalty Effect

According to Reichheld's loyalty effect sustaining customers goes hand in hand with retaining employees and investors and leads to a shifting of investments to most profitable and prospective customers. The stimulation of sustainable growth results in retention of high quality employees, who enrich customer value proposition and hence generate higher growth and profitability. Thus, investors are attracted and tend to higher loyalty, resulting in stabilization, economization, and value-creation. (Reichheld, 1996, pp. 19-25) However, assuming that higher customer retention rates result in increasing profitability is a too simplistic and narrow perspective of the whole process. Customer retention is expected to be reached best by apportioning focus on customers, employees and investors. Substantial strategies in order to retain customers are: (Ahmed, & Buttle, 2002, p. 153; Reichheld, 1996, pp. 64-78)

- Definition and measuring of retention
- Customer quality instead of quantity and the right places to look for their loyalty
- Multiple, changing channels of distribution
- Creative filtering in order to alleviate adverse selection
- Gratifications for continuity and not just new acquisition, through promotional schemes
- Social programs for attraction and retention of the most loyal customers.

Attraction and retention of high-quality employees can be achieved by: (Ahmed, & Buttle, 2002, p. 153; Reichheld, 1996, pp. 64-78)

- Careful recruitment of stuff, who maintain and improve the firm's character and integrity
- Establishment of career paths to improve productivity
- Reward systems not only for acquisition but also for retention of customers
- Adoption of a partnership-concept in order to adjust interests between company and employee.

3.5. Critique in measuring customer retention

Customer retention is often used as a measurement for loyalty, since it is seen as an essential for profitable growth. Nevertheless retention rates, which basically concentrate on customer defections[12], provide valuable linkage to profitability but not to growth rates. High switching costs and other barriers, as well as natural outgrowth of products due to aging, growing income, etc. adulterate retention rates as for customer loyalty. (Reichheld, & Schefter, 2003, p. 4) Since personal treatment of customers and several subjective valuables like loyalty and satisfaction are of high importance for customer retention, it is unassertive if customer retention is even applicable for anonymous mass markets or rather limited to personal contact and relationship marketing. (Dittrich, 2002, p. 2) Furthermore customer satisfaction does not necessarily result in customer retention. (Albers, & Hermann, 2007, p. 850) As shown in different empirical studies of Gierl, 2003 and Reichheld, 1996, numerous customers,

---

[12] "Customer defection is the loss of the whole or a portion of a customer's business", Source: Goeke, McClung, & Reidenbach, 2002, p. 210.

though satisfied defect to competitors. Hence customer satisfaction is an essential, though not effectual requirement for customer retention.

The assumption of many companies, that customers shift their purchase patterns rather than totally stop buying from the company, leads to high managerial focus on the share of wallet and disregard of customer retention rates. Thus, the importance of managing customer retention becomes questionable. (Aksoy, Estrin, Keiningham, & Perkins-Munn, 2005, pp. 245-257)

3.6. Summary

In the 3rd chapter of this thesis the term customer satisfaction was described in detail and various interpretations were discussed, as for example, voluntary and involuntary customer retention. Furthermore opportunities and threats, the effectiveness of customer retention, including the causal chain, critiques on customer retention and managing customer retention were discussed in this chapter. Additionally the steps necessary to establish strategies for retaining customers and possible measurements in the area of customer retention were defined.

Hence the following research questions were answered in this chapter:

- What is customer retention and how to manage it?
- Which opportunities and threats are offered by customer retention?

4. Web 2.0 tools for Customer Retention

In order to be able to evaluate the beforehand described Web 2.0 tools for their applicability to retain customers, different aspects of e-commerce as well as of how to manage customer retention in connection with the respective tools need to be examined.

4.1. Web 2.0 tools in the fields of application of e-commerce

The underlying model of the chapter 4.1. divides the different Web 2.0 tools into four fields of application in the e-commerce sector: product mix policy, pricing policy, communication policy and front-end policy, as shown in table 4.

Table 4: Web 2.0 tools in the fields of application of e-commerce

	Product mix policy	Pricing policy	Communication policy	Front-end policy
Web 2.0 tools	Wikis Subscription service Pod/Video-casts Social Networks (passive)	Social Networks (passive) Social shopping	Subscription service Pod/Video-casts Blogs Social Networks (active)	Tagging Mash-ups

Source: translated and adapted from: Bauer, et al., 2007, p. 205

4.1.1. Product mix policy

The product mix policy includes all decisions regarding the specification of products which are sequentially offered to a company's customers. A wiki offers the possibility to analyze single products regarding their relevance, as well as to add additional information about them in the wiki so that the user-community is able to foster it. The usage of a subscription service as well as pod and video-casts also facilitate the well-directed information processing and furthermore generate user profiles in order to find out about their preferences. Adjustment of the product mix is thus rendered possible by collecting preferred product features and categories. Social networks, though, provide the opportunity of passive monitoring in order to gain evidence about product design, sales pitch, or merely to find out about the user-community's opinion on a product. (Bauer, et al., 2007, pp. 203-207)

4.1.2. Pricing policy

Indicators such as price level, special offers, and especially in the e-commerce sector the usage of electronic coupons or rebate systems, highly depend on a company's pricing policy. The passive usage of social networks once more can be applied through monitoring of the community's preferences. Conditions and price levels of products are often discussed by users of social shopping platforms, which consequently can be used in order to establish a pricing policy. (Bauer, et al., 2007, pp. 204-207)

4.1.3. Communication policy

Decisions about how and what is communicated to customers are determined in the communication policy, which can be supported by newsletter and permission marketing. The utilization of subscription services furthermore fosters this kind of marketing, whereas the usage of social networks should be handled with care. Selective forwarding of information can have an influence on the user-community, though such methods could have negative effects in case it is discovered by the users. Another tool in the communication policy are blogs, since they can be used for commenting and tagging and are hence well functioning communication platforms. (Bauer, et al., 2007, pp. 204-207)

4.1.4. Front-end policy

The front-end policy deals with upfront points of intersection to customers in the e-commerce sector, such as design of the surface, tags or the alignment of a page. With tags customized descriptions and demonstration of contents can be given, and hence a direct point of intersection according to customers' interests is created. Mash-ups use and combine several applications of different providers and hence valorize one's own points of intersection to a customer. (Bauer, et al., 2007, pp. 204-208)

4.2. Facebook vs. Twitter

The online community Facebook is still more popular in German-speaking countries than Twitter for marketing issues, since there are also respectively more German-speaking users at Facebook.[13] Nevertheless Twitter challenges Facebook effectively by means of applications encouraging a developer ecosystem. As a place of innovation and experimentation users consciously or unconsciously are "being fed, counted, pitched, contested, serenaded and otherwise engaged by influencers, power users and marketers who rely on Twitter's developer ecosystem". Though its prospective role as Facebook disruptor, at the beginning of March 2011 Twitter stopped the development of consumer-oriented applications. Despite this cut-back on the developmental side, the creation of enterprise-oriented and

---

[13] See also: Chapter 2.2.3.1. Example

special-purpose applications, as well as the maintenance of existing applications is still guaranteed by Twitter. (Harvard Business School Publishing, March 2011)

Due to this recent events and rumors that, as a consequence, "The Next Big Thing" is about to emerge and anew will prove many experts wrong about the distribution of users' social media hours, this chapter will discuss the applicability of the two Web 2.0 applications Facebook and Twitter for customer retention processes.

4.2.1. Exchange of Information

The exchange of information[14] describes the applicability of the respective tools in order to provide sufficient information to create customer satisfaction and bridge customers' state of mind and their actual behavior.

Facebook can be used in order to spread information about a company and is also useful to create or pursue a certain image. The possibility to create groups can furthermore be used to keep customers up-to-date about recent company events. Despite these possibilities Facebook is often seen as overrun by too many companies and hence users can feel flooded by all the available information. (Fred Benenson's Blog, 2010)

Although direct Tweets are restricted to 140 characters, companies can efficiently use it in order to post and link

---

[14] See also: Chapter 3.3.1.2. Customer satisfaction and Value; 3.4.1. Fundamentals of managing customer retention

interesting articles about their products or special offers, as well as industry specific information. Thus, information can be easily distributed to clients. [15]

4.2.2.   Segmenting and targeting

Managing customer retention requires segmentation and differential targeting of customers in order to determine the right strategy for value creation on buyers' and suppliers' sides.[16]

Due to the huge amount of private user information an efficient instrument for targeting marketing are paid ads, which appear on users' pages according to virtually everything mentioned on their profiles, such as their geographic position or educational background. Such custom-tailored ads, can also lead to increasingly high expenditure if a company's goals are set too high.

Special applications on Twitter are used by companies to automatically contact or follow a targeted group of people. Such applications help marketers e.g. to find people who tweeted something around a certain region, or whose interests are likely for prospective customers. Furthermore involvement with targeted customers can be increased via chats. Twitter lists can also help to segment people according to their profile information or latest updates made. Though all this helpful applications to

---

[15] See also: Chapter 2.2.3.1. Example
[16] See also: Chapter 2.4. Managing customer retention

automate customer service, companies should not rely too much on them, since otherwise users could turn away.

4.2.3. Building trust and acceptance

Customer loyalty highly depends on their trust and acceptance, which in turn are formed by interaction between suppliers and customers and supply of trustworthy contents. [17]

Companies can use pages on Facebook in order to provide their customers with the right information but also to interact with them. Customers feel like a part of a group, with which one can communicate, share interests and preferences. The creation of allegiance and mutual trust is thus fostered.

Twitter can create high interaction with customers through the possibility of commenting and customer service activities. This furthermore creates positive feelings and increases customers' trust and acceptance by publicly dealing with negative issues. CEOs or internal specialists can give their personal recommendations or offer help without an obvious intention to sell and thus, build trust and create a feeling of being cared for. It furthermore offers a very high engagement level and by publishing great content and responding to customers' inquiries, proposals or complaints trust and acceptance can be enhanced. Additionally chats can be used in order to get directly involved with targeted customers.

---

[17] See also: Chapter 3.3.1.3. The loyalty stage

4.2.4. Establishing enthusiasm and emotions

Emotions and enthusiasm heavily influence customers' perception of and attitude towards a company and its products.[18]

Facebook, like most of online communities or social networks[19], is built on emotions, such as a sense of belonging or a corporate feeling, which are perceived by a user. The stronger those emotions are, the more a user is following the community and will continue to do so. Competitions on corporal pages or discussions and participations in groups at Facebook, e.g. where the first 1000 users who click the "like" button win a product sample, can build emotions and also enthusiasm about a company. The negative side can be that groups and pages are over-used for advertisement and thus, people can easily feel spammed by a company and hence create negative feelings.

When reacting to a Tweet, this normally involves emotions or at least an opinion about a certain topic. As a community Twitter also creates a sense of belonging and corporate feeling and hence emotions also play a role. Twitter's advantage is that companies can easily find followers and engage with people even before they are their friends. Thus, emotions can already be created through a very interesting or helpful Tweet at an early stage of the causal chain of customer retention when satisfaction and value are built. Positive feelings can also be established through public handling of negative issues, such as customers' complaints.

---

[18] See also: Chapter 3.3.1.3. The loyalty stage
[19] See also: Chapter 2.2.4. Community and social network

4.2.5. Supporting dialog

The support of dialogs between suppliers and buyers in order to mutually exchange information and opinions is especially important for the creation of customer loyalty. [20]

Companies can support dialog and discussions by creating a group on Facebook. Given that the discussion tab is active every member can start a discussion which can be commented on by any other member of the group. Group discussions need some monitoring and policies in order to be sure not to be overrun by spammers and to keep a friendly atmosphere. Nevertheless the disadvantage of Facebook groups is that discussions can only be joined as a member of this group.

Interesting Tweets can directly lead to interaction and active discussions and dialogs between all Twitter users interested in this topic. Since updates on company information are only read, retwittered or commented on within the first five minutes after being published (KISSmetrics, 2011), promptness and flexibility are an issue. Furthermore every user has the possibility to complain about something or give an opinion on a certain topic. By promptly and directly reacting to such complaints a dialog between supplier and consumer is built. A disadvantage could be that only one user is able to define the main topic.

---

[20] See also: Chapter 3.3.1.3. The loyalty stage

4.2.6. Identifying trends and collecting customer data

The creation of a well-established customer database, including actual customer loss as well as trends and current developments, is an essential for managing and reaching customer retention.[21]

Facebook, as the biggest social network worldwide, commands the biggest user database with loads of voluntarily published personal information. Once users are fan of a company's page or of a group this information can be easily accessed, according to each and everyone's private settings. The huge database of Facebook forms a huge potential for gaining customer information, but is still limited to a user's willingness to share it with companies. Newest trends can be followed on Facebook via popular groups and events which are interrelated with a certain market segment or field of industry.

Twitter offers many applications in order to follow customers' as well as competitors' preferences and actions. The advantage as opposed to Facebook thereby is that this free market research can be conducted before having direct contact with users. Keywords entered by users can be analyzed and used for marketing issues. Nevertheless Twitter does not provide a customer database which is as big as Facebook's.

---

[21] See also: Chapter 3.3.1.4. The final outcome: customer retention

4.3. Summary

As a last chapter of this thesis the applicability of Web 2.0 tools for customer retention was examined in the fields of application in the e-commerce sector. Furthermore two Web 2.0 applications were analyzed into detail as an exemplary, namely Facebook and Twitter. The potential of Web 2.0 tools for customer retention is given and companies investing in this sector can experience improvement. As shown by the two Web 2.0 applications Facebook and Twitter, there is a potential possibility to conduct successful online marketing whenever used the right way.

Hence the following research question was answered in this chapter:

- How can Web 2.0 tools be used in order to successfully manage customer retention?

5. Conclusion

Web 2.0 tools, as defined in chapter 2.2. can be used in all areas of application of e-commerce, some even in more than one at the same time. The four areas, namely product mix, price, communication and front-end policy, are mainly operable in earlier stages of the customer retention process, such as building trust and acceptance, or increasing customer satisfaction and loyalty. A social network, as one of the most useful marketing tools in Web 2.0, is applicable in three of the four respective areas. Thus, a big potential for marketing through Web 2.0 applications such as Facebook, MySpace or Twitter is shown. This versatility is furthermore analyzed in chapter 4.2. by comparing the two social networks Facebook and Twitter.

Both Facebook and Twitter offer potential for retaining customers in different areas such as communication between supplier and customer, the exchange of information and opinions, segmentation and targeting of customers and the identification of trends and collection of customer data.

Whereas Facebook's main advantage is its huge user database and the many different ways of possible marketing, Twitter has the advantage of easy gaining of followers and engagement with people before they are a company's friend or fan. Only until recently Twitter's biggest advantage was its pro-developer stance, with the possibility for everyone to create net applications for marketing use. Since the beginning of March Twitter closed down this developer ecosystem which might be a step towards the creation of a new challenging platform to the existing ones.

Since users at Twitter can choose which content they want to see, traceability of their actions is possible as they act according to their interests. Facebook users on the other hand are swamped by huge amounts of information which are not necessarily in their fields of interests, but still followed by them. With more and more users among the generation 35 – 44 (GigaOM, 2011) young and innovative people might seek for an alternative for spending their social media hours. Especially in German-speaking countries, Twitter has still large potential in user acquisition and hence also future marketing potential for companies is given.

5.1. Outlook to further research

This thesis is solely based on theoretic literature and journals, and includes no empirical studies, best case scenarios or expert opinion from the respective field, thus further need for action is given on research on the practical side. Through interviews and surveys the various tools of customer retention could be established and further illustrated. Furthermore the classification of different Web 2.0 tools for the diverse market sector could be defined in order to guarantee the right access to a targeted customer group. Moreover, the future endurance of examples of Web 2.0 applications, mentioned in chapter 2.2., be examined as a proceeding of this thesis in order to be able to follow and analyze trends in the area of Web 2.0 marketing.

6.   List of References
6.1. Books and Articles in Journals

Ahmad, R., & Buttle, F. (2002). Customer retention management: A reflection of theory and practice. *Marketing Intelligence & Planning, 20, 3, 149-161.*

Aichbauer, S. (2010). *Virales Marketing im Web 2.0 – Chancen und Risiken für Unternehmen.* München: Grin Verlag.

Ajami, R., Gargeya, V., Goddard, G., & Raab, G. (2008). *Customer relationship management: A global perspective.* Hampshire: Gower Publishing Limited.

Aksoy, L., Estrin, D., Keiningham, T., & Perkins-Munn, T. (2005). Actual Purchase as a Proxy for Share of Wallet. *Journal of Service Research, 7, 3, 245-257.*

Albers, S., & Hermann, A. (2007). *Handbuch Produktmanagement.* Wiesbaden: Gabler Verlag.

Alby, T. (2007). *Web 2.0: Konzepte, Anwendungen, Techologien.* München: Carl Hanser Verlag.

Allen, C., Kania, D., & Yaeckel, B. (2001). *One-to-one Web Marketing: Build a Relationship Marketing Strategy One Customer at a Time.* New York: John Wiley & Sons Inc.

Allen, D., & Rao, T. (2000). *Analysis of Customer Satisfaction Data.* Milwaukee: American Society for Quality.

Back, A., Gronau, N., & Tochtermann, K. (2008). *Web 2.0 in der Unternehmenspraxis – Grundlagen, Fallstudien und Trends zum Einsatz von Social Software.* München: Oldenburg Verlag.

Baran, R., Zerres, C., & Zerres, M. (2009). *Customer Relationship Management.* Bookboon.com.

Bauer, H.; Große-Leege, D.; & Rösger, J: (2007). *Interactive Marketing im Web 2.0.* München: Verlag Franz Vahlen

Bliemel, F., & Eggert, A. (1998). Kundenbindung – die neue Sollstrategie?, in: *Marketing Zeitschrift für Forschung und Praxis (ZFP), 1, 37-46.*

Bruhn, M., & Homburg, C. (2005). *Handbuch Kundenbindunsmanagement. Kundenbindungsmanagement – Eine Einführung in die theoretischen und praktischen Problemstellungen.* Wiesbaden: Gabler Verlag.

Chow, S. (2007). *PHP Web 2.0 Mashup Projects.* Birmingham: Packt Publishing Ltd.

Chowdury, A., Jansen, J., Sobel, K., & Zhang, M. (2009). Twitter Power: Twitter as electronic word of mouth. *Journal of the American Society of Information Science and Technology, 60, 9, 1-20.*

Damiani, E., Lytras, M., & Ordonez de Pablos, P. (2009*). Web 2.0: The Business Model.* New York: Springer.

Dawkins, P., & Reichheld, F. (1990). Customer Retention as a Competitive Weapon. *Directors and Boards, 14, 41.*

Diller, H. (1996). Kundenbindung als Marketingziel. *Marketing Zeitschrift für Forschung und Praxis (ZFP), 18, 33-41.*

Dittrich, S. (2002). *Kundenbindung als Kernaufgabe im Marketing: Kundenpotentiale langfristig ausschöpfen.* St. Gallen: Thexis.

Duschinski, H. (2007). *Web 2.0 – Chancen und Risiken für die Unternehmenskommunikation.* Wiesbaden: Diplomica Verlag GmbH.

Finkelman, D., & **Goland**, T. (*1990). How not to Satisfy Your Customers*. The. *McKinsey Quarterly, 4, 2-12.*

Foss, B., & Stone, M. (2001). *Successful Customer Relationship Marketing.* US: Kogan Page.

Gierl, H. (1993). Zufriedene Kunden als Markenwechsel. *Absatzwirtschaft, 36, 2, 90-94.*

Goeke, R., McClung, G., & Reidenbach, E., (2002). *Dominating Markets with Value: Advances in Customer Value Management.* USA: Rhum line Publishing.

Governor, J., Hinchcliffe, D., & Nickull D. (2009). *Web 2.0 Architectures.* Canada: O'Reilly Media, Inc.

Grabenströer, N. (2009). *Web 2.0 – Potenziale im strategischen Marketing.* Köln: Josef Eul Verlag GmbH.

Hagemann, S., & Vossen, G. (2007). *Unleashing Web 2.0 from Concept to Creativity.* USA: Morgan Kaufmann Publishers.

Hass, B., Kilian, T., & Walsh, G. (2008). *Web 2.0: Neue Perspektiven für Marketing und Medien.* Heidelberg: Springer.

Jeremy, T., & Newman, A. (2009). *Enterprise 2.0 Implementation.* New York City: MacGraw-Hill Companies.

Kindermann, H. (2006). *Optimierung der Kundenbindung in Massenmärkten.* Wiesbaden: Deutscher Universitäts-Verlag.

Koch, M., & Richter, A. (2007). *Enterprise 2.0: Planung, Einführung und erfolgreicher Einsatz von Social Software in Unternehmen.* München: Oldenbourg Wissenschaftsverlag.

Meyer, A., & Oevermann, D. (1995). *Kundenbindung.* In: Bruhn, M., & Homburg, C., Handbuch Kundenbindungsmanagement, Wiesbaden: Gabler Verlag.

Mukerjee, K. (2007). *Customer Relationship Management – A strategic approach to Marketing.* New Dehli: Prentice-Hall of India Private ltd.

Müller, W., & Riesenbeck, H. (1991). Wie aus zufriedenen Kunden auch anhängliche Kunden werden. *Harvard Manager, 3, 68.*

Newman, A., Thomas, J. (2009). *Enterprise 2.0 implementation.* New York: The MacGraw-Hill Companies.

Reichheld, F. (1996). *The Loyalty Effect.* Boston: Harvard Business School.

Sauer, M. (2006). *Weblogs, Podcasting & Online-Journalismus.* Köln: O'Reilly Verlag.

Schrum L., Solomon G. (2007). *Web 2.0 – New tools, new schools. Washington* DC: International Society for Technology in Education.

Shuen, A. (2008). *Web 2.0: A strategy guide.* Canada: O'Reilly Media, Inc.

Stauffer, T. (2008). *How to do Everything with Your Web 2.0 Blog.* New York: The MacGraw-Hill Companies.

Töpfer, A. (1996). *Kundenzufriedenheit messen und steigern.* Neuwied: BBE Verlag.

Völtz, G. (2011). *Die Werkwiedergabe im Web 2.0.* Wiesbaden: Gabler Verlag.

Weinberg, T. (2010). *Social Media Marketing – Strategien für Twitter, Facebook, & Co.* Köln: O'Reilly Verlag.

Werani, T. (2003). *Bewertung von Kundenbindungsstrategien in B-to-B Märkten.* Wiesbaden: Deutscher Universitäts-Verlag.

6.2. Online Journals and Documents

Alf Nucifora (1997-2009). *Alf's Articles: Sales and Marketing Win big With Online Gaming.* Downloaded on April 3rd, 2011, from: http://www.nucifora.com/art_218.html.

AOL Inc. (2011). *Yahoo is shutting down Delicious, heralding the end of a tasty era.* Downloaded on March 27th, 2011, from:

http://downloadsquad.switched.com/2010/12/16/yahoo-shutting-down-delicious/.

Barracuda Networks, Inc. (2003-2011). *Twitter Makes its Red Carpet Debut for Celebrities and Criminals.* Downloaded on March 27th, 2011, from: http://www.barracudanetworks.com/ns/news_and_events/index.php?n id=387

Bob Mobile AG (2011). *Corporate News.* Downloaded on April 10th, 2011, from: http://www.bobmobile.ag/pdf/corporate_de/2011_02_01_CN_Bob_Mo bile_Gondal_D.pdf?PHPSESSID=093dd1d5c2fc70ab0fa93374d101fce8.

Boutell.com (1994-2011). *WWW FAQs: Why is myspace so popular and what is it good for?* Downloaded on April 10th, 2011, from: http://www.boutell.com/newfaq/sitespecific/whymyspace.html.

Bulk Email marketing Tips & Services (2009). *Key message copy platforms: the Secret to Increasing the Selling Power of your B2b Marketing Materials.* Downloaded on April 10th, 2011, from: http://www.hellomails.com/blog/?p=21128.

Business E-Coaching (2010). *Customer Retention.* Downloaded on Jannuary 30th, 2011, from: http://www.1000ventures.com/business_guide/crosscuttings/customer _retention.html.

Customer Experience Labs (2008). *The confirmation/disconfirmation paradigm: why satisfied customers are not always satisfied.* Downloaded on February 5th, 2011, from: http://www.customer-experience-labs.com/2008/04/28/the-confirmationdisconfirmation-paradigm-why-satisfied-customers-are-not-always-satisfied/.

derStandard.at (2011). *Facebook übernimmt Foto-Plattform.* Downloaded on March 27th, 2011, from: http://derstandard.at/1269448858508/Facebook-uebernimmt-Foto-Plattform.

Digg.com (2011). *Digg it!* Downloaded on April 10th, from: http://digg.com/news/technology/digg_it_6.

Digital Buzz Blog (2008). *Facebook: Facts and Figures for 2010.* Downloaded on March 19th, 2011, from: http://www.digitalbuzzblog.com/facebook-statistics-facts-figures-for-2010/.

eBizMBA, Inc. (2011). *Top 15 Most Popular Social Networking Websites – April 2011.* Downloaded on April 10th, 2011, from: http://www.ebizmba.com/articles/social-networking-websites.

EzineArticles.com (2011). *History of Online Games.* Downloaded on April 3rd, 2011, from: http://ezinearticles.com/?History-of-Online-Games&id=1649861.

Flickr (2011). *All time most popular tags.* Downloaded on March 19th, 2011, from: http://www.flickr.com/photos/tags/.

Flimp Media (2011). *Online Video Platform Marketing and Communications.* Downloaded on March 27th, 2011, from: http://www.flimp.net/video-marketing-platform.php.

Fred Benenson's Blog (2010). *Information Overload, Facebook Fatigue and Twitter's Awesome Filter.* Downloaded on April 3rd, 2011, from: http://fredbenenson.com/blog/2009/01/02/information-overload-facebook-fatigue-and-twitters-awesome-filter/.

GigaOM (2011). *Facebook vs. Twitter: An Infographic.* Downloaded on April 3rd, 2011, from: http://gigaom.com/2010/12/20/facebook-vs-twitter-an-infographic/.

Harvard Business School Publishing (2011). *Playing Games with Customers.* Downloaded on April 3rd, 2011, from: http://hbr.org/2003/04/playing-games-with-customers/ar/1.

Harvard Business School Publishing (2010). *Twitter Locks Down, ending its Reign as the Next Big Thing.* Downloaded on April 1st, 2011, from: http://blogs.hbr.org/samuel/2011/03/twitter-locks-down-ending-its.html.

Hub Spot, Inc. (2011). *The ultimate list: 100+ Twitter Statistics.* Downloaded on March 27th, 2011, from:

http://blog.hubspot.com/blog/tabid/6307/bid/6050/The-Ultimate-List-100-Twitter-Statistics.aspx.

Huffington Post (2011). *The Huffington Post.* Downloaded on April, 10th, 2011, from: http://www.huffingtonpost.com/.

Illustree (2007). *Online und Marketing-Games.* Downloaded on April 3rd, 2011, from:
http://www.illustree.at/de/services/onlinemarketing/online_games.

Internet Alchemy (2011). *Talis, Web 2.0 and All That.* Downloaded on March 13th, 2011, from: http://blog.iandavis.com/2005/07/04/talis-web-2-0-and-all-that/.

Internetberatung Sven Przepiorka (1998-2011). *Tagging – Schlagwörter erobern das Internet.* Downloaded on April 10th, 2011, from: http://tzwaen.com/publikationen/tagging-schlagwoerter/.

Internet und Werbeagentur Dresden (1999-2010). *Online Games/Simulationen.* Downloaded on April 10th, 2011, from: http://www.dtele.de/leistungen/online-games-simulationen/.

KISSmetrics (2011). *Facebook Marketing: A Comprehensive Guide for Beginners* Downloaded on April 3rd, 2011, from: http://blog.kissmetrics.com/facebook-marketing/.

KISSmetrics (2011). *Twitter Marketing Guide.* Downloaded on April 3rd, 2011, from: http://blog.kissmetrics.com/twitter-marketing-guide/.

Learning Theories (2008). *Communities of Practice.* Downloaded onMarch 13th, 2011, from: http://www.learning-theories.com/communities-of-practice-lave-and-wenger.html.

Linden Research, Inc., (2011). *Second Life.* Downloaded on April 10th, 2011, from: http://secondlife.com/.

Marktstudie (2011). *Kundenbindungsmanagement.* Downloaded on Jannuary 30th, 2011, from: http://www.markt-studie.de/studien/Kundenbindung%20und%20Kundenrueckgewinnung.php.

MetaFilter Network, Inc. (1999-2011). *What is the best Photo blogging Platform?* Downloaded on April 10th, 2011, from: http://ask.metafilter.com/169433/What-is-the-best-photo-blogging-platform.

Mister Wong (2011). *Freie Bibliothek digitaler Dokumente.* Downloaded on April 10th, 2011, from: http://www.mister-wong.de/.

New York Times Company (2011). *RSS.* Downloaded on March 19th, 2011, from: http://www.nytimes.com/services/xml/rss/index.html.

NevOn (2004-2006). *Blogs vs. forum: What's the difference?* Downloaded on March 20th, 2011, from: http://www.nevon.net/nevon/2004/10/blog_vs_forum_w.html.

Nick Tadd (2011). *Ning to Charge Users.* Downloaded on April 10th, 2011, from: http://propertytribes.ning.com/forum/topics/ning-to-charge-users.

Ning, Inc. (2010). *The main reason why members are not coming back to your site.* Downloaded on April 10th, 2011, from: http://creators.ning.com/forum/topics/the-main-reason-why-members?id=4244211%3ATopic%3A466983&page=3.

O'Reilly Media Inc. (2011). *What is Web 2.0.* Downloaded on February 14th, 2011, from: http://oreilly.com/web2/archive/what-is-web-20.html.

O'Reilly radar (2005-2009). *Not 2.0?* Downloaded on February 14th, 2011, from: http://radar.oreilly.com/archives/2005/08/not-20.html.

Österreichische Akademie der Wissenschaft. (2011). *Microblogging und die Wissenschaft.* Downloaded on March 27th, 2011, from: http://epub.oeaw.ac.at/ita/ita-projektberichte/d2-2a52-4.pdf.

Podcast.at (2010). *Podcast Neuzugänge.* Downloaded on April 10th, 2011, from: http://www.podcast.at/podcasts.html.

Reicheld, F., & Schefter, P. (2003). *The One Number You Need To Grow. Harvard Business Review.* Downloaded on January 9th, 2011 from: http://www.netzkobold.com/uploads/pdfs/the_one_number_you_need _to_grow_reichheld.pdf.

Schmidt, H. (2009). *Die Angst der Unternehmen vor Twitter.* Downloaded on March 28th, 2011, from: http://faz-community.faz.net/blogs/netzkonom/archive/2009/05/11/twitter-fuer-unternehmen.aspx.

SideGround, Inc., (2011). *Best Wiki: Comparison of the best free wiki software.* Downloaded on April 10th, 2011, from: http://www.siteground.com/compare_best_wiki.htm.

SlideShare Inc. (2011). *Social Bookmark Marketing with del.icio.us.* Downloaded on March 28th, 2011, from: http://www.slideshare.net/tunheimpartners/social-bookmark-marketing-with-delicious.

Square Jaw Media (March, 2011). *Beyond the Basics: Corporate Microblogging.* Downloaded on March 27th, 2011, from: http://www.squarejawmedia.com/2011/03/beyond-the-basics-corporate-microblogging.html.

Technorati, Inc. (2011). *Top 100 Blogs – 1-25.* Downloaded on April 10th, 2011, from: http://technorati.com/blogs/top100.

WebBizIdea.com (2011). *How LinkedIn works & why it is so popular.* Downloaded on April 10th, 2011, from: http://blog.webbizideas.com/how-linkedin-works-why-it-is-so-popular/.

WebResourcesDepot (2011). *9 Open Source Microblogging Applications.* Downloaded on April 10th, 2011, from: http://www.webresourcesdepot.com/9-open-source-microblogging-applications/.

Wikipedia (2011). *Wikipedia:Etiquette.* Downloaded on March 28th, 2011, from: http://en.wikipedia.org/wiki/Wikipedia:Etiquette.

www.IHans.de (2003-2008). *Zusammenfassung der besten Videoplattformen im Internet.* Downloaded on April, 10th, 2011, from:

http://www.ihans.de/newsdetails/datum/2007/09/14/zusammenfassun
g-der-besten-video.html.

Yahoo! Inc. (2011). *Why is Twitter so popular?* Downloaded on April 10th, 2011,
from:
http://answers.yahoo.com/question/index?qid=20090419162100AA1Zk
7q.

Ziff Davis, Inc. (1996-2011). *Definition of API.* Downloaded on March 21st, 2011
from:
http://www.pcmag.com/encyclopedia_term/0,2542,t=application+progr
amming+interface&i=37856,00.asp.

www.ingramcontent.com/pod-product-compliance
Lightning Source LLC
Chambersburg PA
CBHW070911180526
45168CB00005B/2001